YOUR KNOWLEDGE HAS VALUE

- We will publish your bachelor's and master's thesis, essays and papers

- Your own eBook and book - sold worldwide in all relevant shops

- Earn money with each sale

Upload your text at www.GRIN.com and publish for free

Bibliographic information published by the German National Library:

The German National Library lists this publication in the National Bibliography;
detailed bibliographic data are available on the Internet at http://dnb.dnb.de .

Imprint:

Copyright © 2016 GRIN Verlag, Open Publishing GmbH
Print and binding: Books on Demand GmbH, Norderstedt Germany
ISBN: 9783668342576

This book at GRIN:

http://www.grin.com/en/e-book/343915/on-the-conservation-of-momentum-
angular-momentum-energy-and-information

Alexander Mircescu

On the Conservation of Momentum, Angular Momentum, Energy, and Information

GRIN Publishing

On the Conservation of Momentum, Angular Momentum, Energy, and Information

Dr. Alexander Mircescu, Munich, Germany

1. Force F and Torque T

1.1 Force F

The force $\vec{F}$ is, generally spoken, a vector function of space and time [1]. We have:

$$\vec{F}(\vec{r}, t)$$

$\vec{F}$ describing the magnitude and direction of the force vector, $\vec{r}$ describing a vector in three-dimensional space (x, y, z), and t describing a point in time.

The force $\vec{F}$ can be applied in a jump manner to a particle or a body, see figure 1. Hence, the force function does not allow a conservation law.

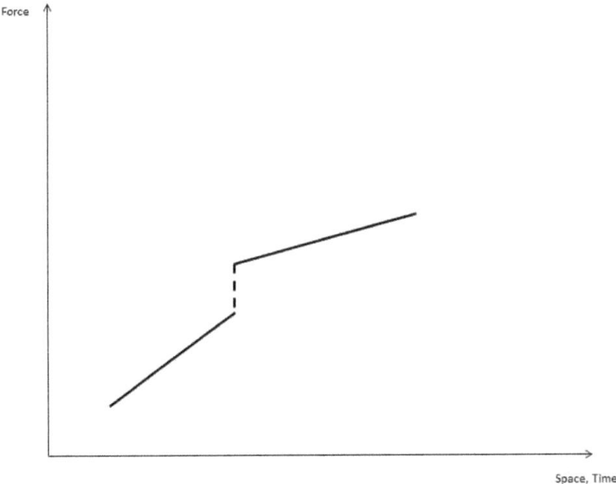

Figure 1: Force in Space and Time

The force $\vec{F}$ is defined by the units $[\vec{F}] = N = \frac{kg\,m}{s^2}$, according to the second law of Newton's laws of motion [1].

1.2 Torque T

The torque $\vec{T}(\vec{\varphi}, t) = \vec{r} \times \vec{F}(\vec{r}, t)$, (with $\vec{F}(\vec{r}, t)$ as force, $\times$ as cross product, and $\vec{r}$ as position vector for the force) [1] is, generally spoken, a vector function of angle $\vec{\varphi}$ and time t. We have:

$$\vec{T}(\vec{\varphi}, t)$$

$\vec{T}$ describing the magnitude and direction of the torque vector, $\vec{\varphi}$ describing an angle vector in three-dimensional space (x, y, z), and t describing a point in time.

The torque $\vec{T}$ can be applied in a jump manner to a particle or a body, see figure 2. Hence, the torque function does not allow a conservation law.

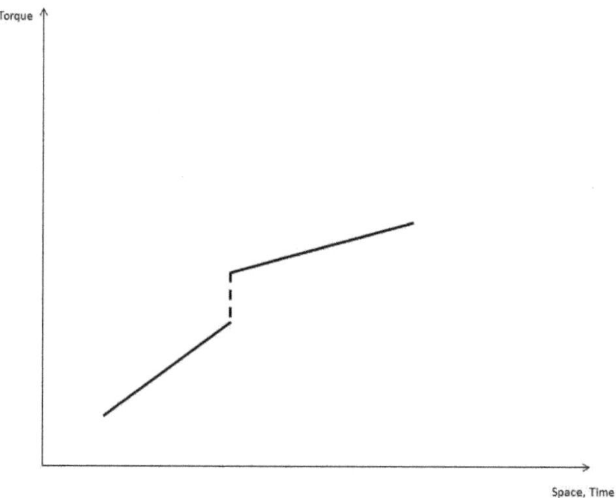

Figure 2: Torque in Space and Time

The torque $\vec{T}$ is hence defined by the units $[\vec{T}] = N\,m = \frac{kg\,m^2}{s^2}$.

2. Momentum p, Angular Momentum L, Overall Momentum, and Energy E

2.1 Momentum p

We can integrate the force in time in order to obtain the momentum [1]:

$$\int_{t_1}^{t_2} \vec{F}(\vec{r}, t)\, dt = \vec{p}(\vec{r})$$

The momentum $\vec{p}$ is a vector function which is dependent of space $\vec{r}$ but not longer of time t, since the time dependence has disappeared by integrating the force $\vec{F}$ in the time interval $\Delta t = t_2 - t_1$.

Due to the integration of a jump in the force function, the momentum function has no jump any more, see figure 3. Hence, the momentum function allows a conservation law.

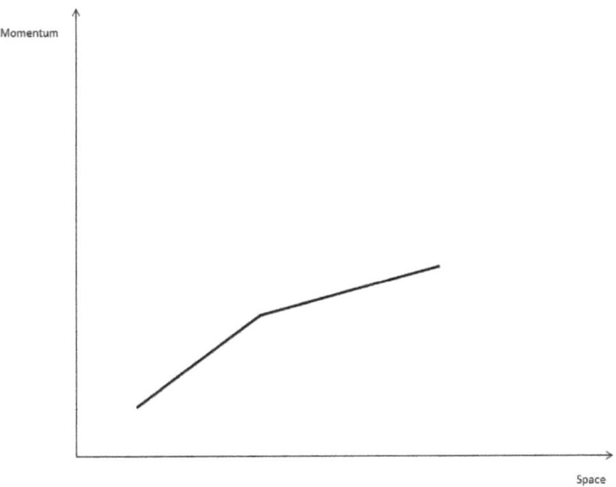

Figure 3: Momentum in Space

One fundamental principle of physics is that the momentum $\vec{p}$ is conserved in a closed system. This can be derived from Newton's laws of motion [1].

Hence, in a closed system we have [1]:

$$\vec{p}_1 = \vec{p}_2 \;\; (\boldsymbol{Conservation\ of\ momentum})$$

$\vec{p}_1$ describing the momentum before the interaction, and $\vec{p}_2$ after the interaction.

The momentum $\vec{p}$ is hence defined by the units $[\vec{p}] = N\,s = \dfrac{kg\,m}{s}$.

2.2 Angular Momentum L

When we turn from translational to rotational movements, 5 units have to be mapped [1], see table 1.

Translational Unit	Rotational Unit
$Length\ \vec{l}$ $[\vec{l}] = m$	$Rotation\ Angle\ \vec{\varphi}$ $[\vec{\varphi}] = rad,\ hence\ dimensionless$
$Velocity\ \vec{v} = \dfrac{\overrightarrow{dl}}{dt}$ $[\vec{v}] = \dfrac{m}{s}$	$Angular\ velocity\ \vec{\omega} = \dfrac{\overrightarrow{d\varphi}}{dt}$ $[\vec{\omega}] = \dfrac{1}{s}$
$Acceleration\ \vec{a} = \dfrac{\overrightarrow{dv}}{dt}$ $[\vec{a}] = \dfrac{m}{s^2}$	$Angular\ acceleration\ \vec{\alpha} = \dfrac{\overrightarrow{d\omega}}{dt}$ $[\vec{\alpha}] = \dfrac{1}{s^2}$
$Force\ \vec{F}$ $[\vec{F}] = N$	$Torque\ \vec{T} = \vec{r} \times \vec{F}$ $[\vec{T}] = N\,m$
$Mass\ m$ $[m] = kg$	$Moment\ of\ Inertia\ M = \displaystyle\int_{m_s}^{m_e} r^2 dm$ $[M] = m^2\,kg$

Table 1: Translational and Rotational Units ($\vec{r}$ defines the position vector of the force $\vec{F}$; m_s and m_e define the starting and ending points, respectively, when integrating the mass, m, of the body; r defines the distance of the body from the rotation axis.)

We can integrate the torque in time in order to obtain the angular momentum [1]:

$$\int_{t_1}^{t_2} \vec{T}(\vec{\varphi}, t)\,dt = \vec{L}(\vec{\varphi})$$

The angular momentum $\vec{L}$ is a vector function which is dependent of the angle $\vec{\varphi}$ but not longer of time t, since the time dependence has disappeared by integrating the torque $\vec{T}$ in the time interval $\Delta t = t_2 - t_1$.

Due to the integration of a jump in the torque function, the angular momentum function has no jump any more, see figure 4. Hence, the angular momentum function allows a conservation law.

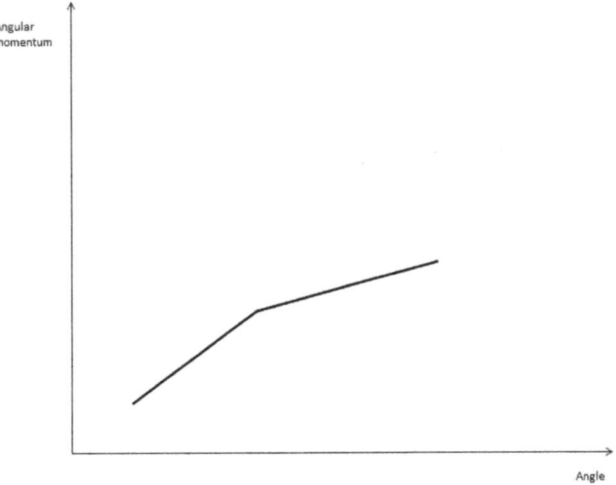

Figure 4: Angular Momentum in the Angle $\vec{\varphi}$

One fundamental principle of physics is that the angular momentum $\vec{L}$ is conserved in a closed system. This can be derived from Newton's laws of motion [1].

Hence, in a closed system we have [1]:

$$\vec{L_1} = \vec{L_2} \; (\textbf{\textit{Conservation of angular momentum}})$$

$\vec{L_1}$ describing the angular momentum before the interaction, and $\vec{L_2}$ after the interaction.

The angular momentum $\vec{L}$ is hence defined by the units $\left[\vec{L}\right] = N\,m\,s = \frac{kg\,m^2}{s}$.

2.3 Overall Momentum

The momentum $\vec{p}$ describes the translational movements; the angular momentum $\vec{L}$ describes the rotational movements. Said two momenta superpose to the overall momentum [1].

2.4 Energy E

We can integrate the force in space and the torque in the angle $\vec{\varphi}$ in order to obtain the energy [1]:

$$\int_{\vec{r}_1}^{\vec{r}_2} \vec{F}(\vec{r},t)d\vec{r} + \int_{\vec{\varphi}_1}^{\vec{\varphi}_2} \vec{T}(\vec{\varphi},t)d\vec{\varphi} = E(t)$$

The energy E is a scalar function which is dependent of time t but not longer of space $\vec{r}$ and on the angle $\vec{\varphi}$, since the space dependence has disappeared by integrating the force $\vec{F}$ in the space interval $\Delta\vec{r} = \vec{r}_2 - \vec{r}_1$, and since the dependence on the angle $\vec{\varphi}$ has disappeared by integrating the torque $\vec{T}$ in the angle interval $\Delta\vec{\varphi} = \vec{\varphi}_2 - \vec{\varphi}_1$.

Due to the integration of a jump in the force and torque functions, the energy function has no jump any more, see figure 5. Hence, the energy function allows a conservation law.

One fundamental principle of physics is that the energy E is conserved [1]. Hence, we always have

$$E_1 = E_2 \ (\boldsymbol{Conservation\ of\ energy})$$

E_1 describing the energy before the interaction, and E_2 after the interaction.

The energy E is defined by the units $[E] = J = N\,m = \frac{kg\,m^2}{s^2}$.

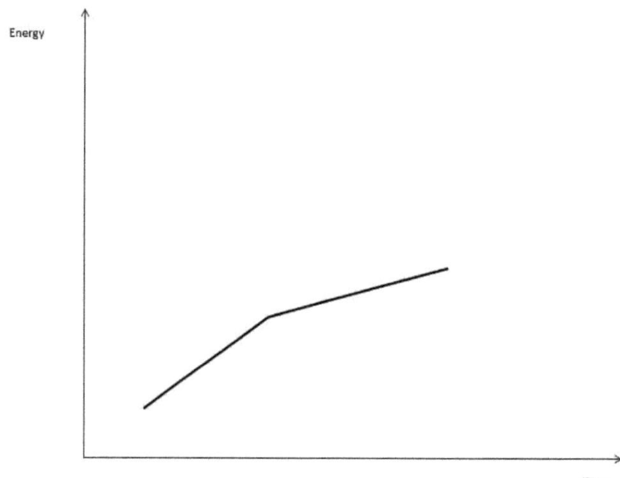

Figure 5: Energy in Time

3. Information I

We can integrate the force in space and time and the torque in the angle $\vec{\varphi}$ and in time, or we can integrate the momentum in space and the angular momentum in the angle $\vec{\varphi}$, or we can integrate the energy in time to obtain the information I:

$$\iint_{t_1 \vec{r}_1}^{t_2 \vec{r}_2} \vec{F}(\vec{r},t)\,d\vec{r}\,dt + \iint_{t_1 \vec{\varphi}_1}^{t_2 \vec{\varphi}_2} \vec{T}(\vec{\varphi},t)\,d\vec{\varphi}\,dt =$$

$$\int_{\vec{r}_1}^{\vec{r}_2} \vec{p}(\vec{r})\,d\vec{r} + \int_{\vec{\varphi}_1}^{\vec{\varphi}_2} \vec{L}(\vec{\varphi})\,d\vec{\varphi} =$$

$$\int_{t_1}^{t_2} E(t)\,dt =$$

$$I$$

The information I is a scalar which is neither dependent on space nor on time, since the space dependence has disappeared by integrating the force $\vec{F}$ in the space interval $\Delta\vec{r} = \vec{r}_2 - \vec{r}_1$ and the torque $\vec{T}$ in the angle interval $\Delta\vec{\varphi} = \vec{\varphi}_2 - \vec{\varphi}_1$, and since the time dependence has disappeared by integrating the force $\vec{F}$ and the torque $\vec{T}$ in the time interval $\Delta t = t_2 - t_1$.

Hence, information I is described by a scalar (by a number) and not by a function, like the force, momentum, angular momentum, or energy.

I summarizes the momentum of the space interval $\Delta\vec{r} = \vec{r}_2 - \vec{r}_1$, the angular momentum of the angle interval $\Delta\vec{\varphi} = \vec{\varphi}_2 - \vec{\varphi}_1$, and the energy of the time interval $\Delta t = t_2 - t_1$ in an index I.

The information I is defined by the units $[I] = J\,s = N\,m\,s = \dfrac{kg\,m^2}{s} := bit$.

Corollary 1: Hence, we have defined the information I in a physical manner by using a force and torque applied to a particle/body, and have connected this information to bits used in information technology.

Corollary 2: Since information is not defined as a function of space and/or time but by a number, we cannot formulate a law of conservation of information at this stage, since we cannot speak of time or space before and after the interaction. Instead, I summarizes the momentum, the angular momentum, and the energy of the interaction process and represents the result as a number defining bits.

4. Planck units

In physics, we can use five universal constants (the Planck constants) in order to define units of measurement [2].

The gravitational constant, G; the speed of light in vacuum, c; the Planck constant, h; the Coulomb constant, $\dfrac{1}{4\pi\varepsilon_0}$; the Boltzmann constant k_B. Since we are interested in the force, torque, momentum, angular momentum, energy, and information, but not in electrical and thermodynamic processes, we focus on the first three Planck units.

The gravitational constant, G, is important when defining the force between two particles/bodies [1]:

$$F = G\frac{m_1 m_2}{r^2} \quad (1): Matter\ equation$$

m_1 describing the mass of the first particle/body; m_2 describing the mass of the second particle/body; and r describing the radius between the first and second particle/body. The Force applied to matter is the key element in the introduction of momentum, angular momentum, energy, and information. Hence, equation (1) shall be defined the **matter equation**.

The speed of light in vacuum, c, is important when defining the relationship between matter (given by the mass m) and energy, E [1]:

$$E = mc^2 \quad (2): Energy\ equation$$

Hence, equation (2) shall be defined the **energy equation**.

The Planck constant, h, is important when defining the relationship between energy, E, and frequency/time [1]:

$$E = hf \quad (3): Information\ equation$$

E defines the energy of the particle/wave and f defines its frequency. We have $f = \frac{1}{T}$, with T being the period of the wave. The Planck constant, h, has the same dimension as information: $[I] = [h] = J\ s = N\ m\ s = \frac{kg\ m^2}{s}$. Indeed, the equations

$$\Delta p_x \Delta x \approx h \quad and \quad \Delta E \Delta t \approx h$$

known as Heisenberg's uncertainty principle [1] state that the information defined by $\Delta p_x \Delta x$ or $\Delta E \Delta t$ is at least h. Hence, h is the smallest possible information. Hence, we shall call equation (3) as the **information equation**.

5. Petri nets

As already mentioned in corollary 2, I summarizes the momentum, angular momentum, and the energy of the interaction process and represents the result as a number defining bits. This fact can be easily depicted by Petri nets. Petri nets are defined in [3]. A possible Petri net is shown in figure 6.

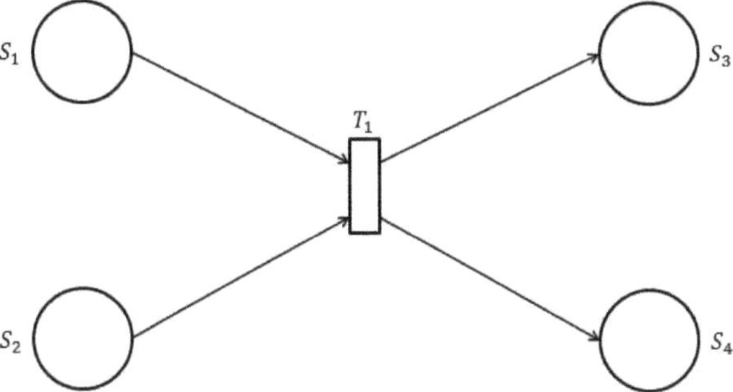

Figure 6: Petri Net

The circles $S_1 - S_4$ are called places and represent the state of a system. We can identify the places with information as described above. Hence, the places $S_1 - S_4$ identify four information pieces $I_1 - I_4$.

The rectangle T_1 is called transition and represents, together with the arrows, the causal relationship between the Information pieces.

Corollary 3: Hence, Petri nets are causal nets representing the causal relationship of information pieces.

Corollary 4: Information as defined above cannot be shown in space and/or time, since said information is not a function of space and/or time. But said information can be shown in a causal net.

6. Physical measurements

It is a well known principle of quantum theory that a measurement disturbs the system which is measured [1]. Such an example is shown in figure 7.

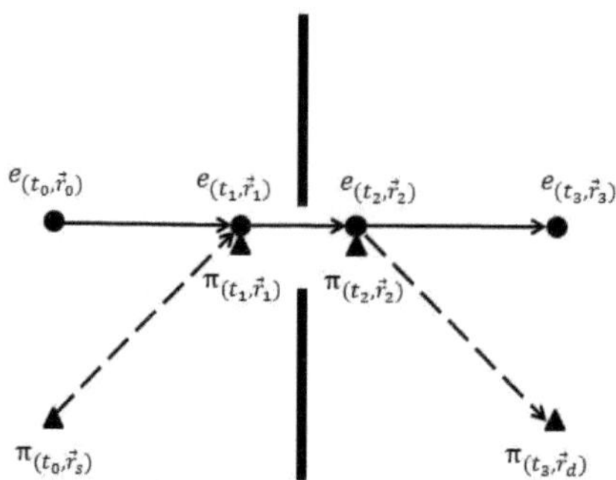

Figure 7: Physical Measurement

The triangle represents a photon π used to perform a measurement, and the circle represents an electron e which is measured by the photon π. The photon starts at time t_0, arrives at time t_1 at the electron, interacts with the electron from time t_1 to time t_2, is scattered from the electron at time t_2, and arrives at the observer at time t_3. The same applies for the spatial coordinates r_i, of course, and needs not to be repeated.

The interaction between electron and photon is guided, of course, by the basic laws of physics like the law of conservation of energy, and the law of conservation of momentum and angular momentum (see the discussion above).

We have before the interaction, hence before the time t_1:

$$E_{bf} = E_1 + E_2$$

E_1 describing the energy of the photon before the interaction, and E_2 describing the energy of the electron before the interaction.

We have after the interaction, hence after the time t_2:

$$E_{af} = E_3 + E_4$$

E_3 describing the energy of the photon after the interaction, and E_4 describing the energy of the electron after the interaction.

According to the law of conservation of energy we have:

$$E_1 + E_2 = E_3 + E_4 \quad (*)$$

The energy $E_1 + E_2$ is transformed during the interaction (hence between t_1 and t_2) by an energy flow to the energy $E_3 + E_4$; and the energy $E_3 + E_4$ is set up during the interaction (hence between t_1 and t_2) by said energy flow from the energy $E_1 + E_2$. Hence, we can integrate equation $(*)$ in time for the time period of interaction (from t_1 to t_2), and obtain:

$$\int_{t_1}^{t_2} E_1 dt + \int_{t_1}^{t_2} E_2 dt = \int_{t_1}^{t_2} E_3 dt + \int_{t_1}^{t_2} E_4 dt \quad (**)$$

Equation $(**)$ can be written as:

$$I_1 + I_2 = I_3 + I_4 \quad (\boldsymbol{Conservation\ of\ information})$$

Remark: We could have derived the law of conservation of information also by integrating the momentum in space and the angular momentum in the angle $\vec{\varphi}$, and by using the law of conservation of momentum and angular momentum. We could also have derived the law of conservation of information by integrating the force in space and time and the torque in the angle $\vec{\varphi}$ and in time, and by using the law of conservation of momentum, angular momentum, and energy.

Corollary 5: The validity of the laws for conservation of energy, momentum, and angular momentum leads to the law of conservation of information. No information gets lost during a measurement.

Corollary 6: Whereas the laws of conservation of energy, momentum, and angular momentum can be directly observed in the local reference frame of the interacting particles, the law of conservation of information can only be observed during a measurement by using the local reference frame of the particles and the local reference frame of the observer.

Corollary 7: According to Albert Einsteins's special theory of relativity the reference frames of particles and observer are connected by a Lorentz transformation [1]. Hence, space and time get transformed from one reference frame to another, and the momentum, angular momentum, and energy also get transformed between both reference frames [1]. Due to $(*)$ and $(**)$ the information I gets transformed between both reference frames connected by a Lorentz transformation.

Corollary 8: Information cannot be observed in a spacetime diagram like momentum, angular momentum, and energy, since I is not a function of spacetime. But information can be observed in a causal net. Since the causal net consists of two dimensions (places and transitions) [3], the observation of information adds two new dimensions.

7. Information and causality

As already discussed, information is naturally depicted by causal nets. A causal net possesses two distinct elements: the places (represented by circles) and the transitions (represented by rectangles). Figure 8 shows the two possibilities of an elementary causal net [3].

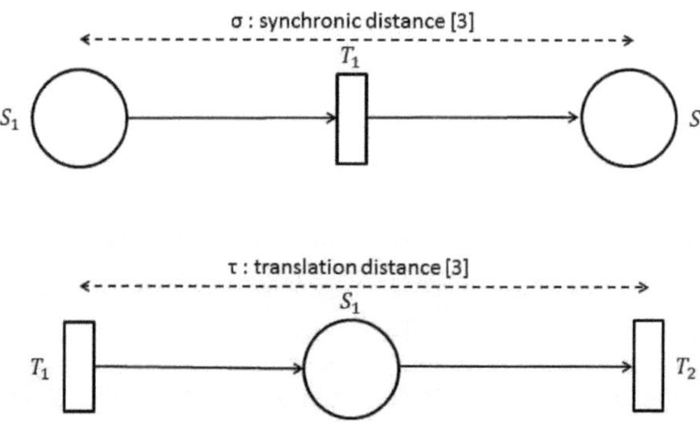

Figure 8: The two Elementary Structures of a Causal Net

As shown in [3], every place S has to be connected to a transition T but not to another place; and every transition T has to be connected to a place S but not to another transition. Therefore, the two possibilities $S_1 - T_1 - S_2$ or $T_1 - S_1 - T_2$ define the elementary net structures. It is proven in [3] that the places S and the transitions T are dual entities, and that a causal net defines a continuum.

Carl Adam Petri shows in [3] that one can define a synchronic distance σ between the places, and a translation distance τ between the transitions. The longer the distances σ and τ are, the more places and transitions, respectively, are crossed. Hence, σ and τ define two possible axes, one for the places and one for the transitions, respectively. In [3] σ and τ are dimensionless, since they are applied to the abstract structure of causal nets.

We have already shown in corollary 3 that we can identify the places with information pieces I_i. In this case, the synchronic distance σ defines an information axis, I, defining distances among the information pieces I_i, bearing in mind that several information pieces can occupy the same location on the information axis defined by σ. In the case of information, σ is not longer dimensionless (like for abstract causal nets), but possesses the dimension of information, namely $[I] = J\,s = N\,m\,s = \frac{kg\,m^2}{s}$ $:= bit$.

The transitions define the interaction of the information pieces I_i according to equation (∗∗) above. Hence, we have information pieces $I_1 \dots I_n$ before the interaction, such that said information pieces are connected by the transition to information pieces $I_1 \dots I_m$ after the interaction, and such that $\sum_{i=1}^{i=n} I_i = \sum_{i=1}^{i=m} I_i$ due to the law of conservation of information.

Otherwise stated, we have a vector of information pieces before the interaction: $\vec{I_S} = \begin{pmatrix} I_1 \\ \vdots \\ I_n \end{pmatrix}$, and we have a vector of information pieces after the interaction: $\vec{I_D} = \begin{pmatrix} I_1 \\ \vdots \\ I_m \end{pmatrix}$. The transition acts as a matrix $\underline{K}$ projecting $\vec{I_S}$ to $\vec{I_D}$. The matrix $\underline{K}$ has hence the structure $\underline{K} = \begin{pmatrix} k_{11} & \cdots & k_{1n} \\ \vdots & \ddots & \vdots \\ k_{m1} & \cdots & k_{mn} \end{pmatrix}$.

Hence, we have the equation

$$\begin{pmatrix} I_1 \\ \vdots \\ I_m \end{pmatrix} = \begin{pmatrix} k_{11} & \cdots & k_{1n} \\ \vdots & \ddots & \vdots \\ k_{m1} & \cdots & k_{mn} \end{pmatrix} \cdot \begin{pmatrix} I_1 \\ \vdots \\ I_n \end{pmatrix}$$

showing how the information before the interaction is projected on the information after the interaction. Hence, the matrix $\underline{K}$ showing the interaction described by a transition is dimensionless. We identify the matrix $\underline{K}$ with causality, and conclude that causality is dimensionless, contrary to information.

We can use the translation distance τ to define a causality axis, k, defining distances among the matrices $\underline{K_i}$, bearing in mind that several matrices can occupy the same location on the causality axis defined by τ. In the case of causality, τ remains dimensionless like for abstract causal nets.

The dimensions for information I and for causality k are the two additional dimensions mentioned in corollary 8.

8. Joint consideration of momentum, angular momentum, energy, and information in space, time, and causality

In physics, the conservation of momentum, angular momentum and energy are fundamental principles as discussed above. Furthermore, we have derived a conservation of information in this paper. Hence, we have four conservation laws:

$$\vec{p}_1 = \vec{p}_2 \; (\textbf{\textit{Conservation of momentum}})$$

$$\vec{L}_1 = \vec{L}_2 \; (\textbf{\textit{Conservation of angular momentum}})$$

$$E_1 = E_2 \; (\textbf{\textit{Conservation of energy}})$$

$$I_1 = I_2 \; (\textbf{\textit{Conservation of information}})$$

Corollary 9: Since momentum and angular momentum are quantities describing the features of particles, momentum and angular momentum can be viewed as features of matter. Therefore, we have conservation laws describing how matter, energy, and information are conserved.

In quantum theory, Werner Heisenberg has shown that the uncertainty principle is valid when measuring the momentum range Δp_x and spatial range Δx of a particle, or the energy range ΔE and the temporal location range Δt of a particle. The same applies, of course, for the measurement of the angular momentum range ΔL_{xy} of a particle and the angle location range $\Delta \varphi_{xy}$ of said particle.

Carl Adam Petri shows in [3] that the causal nets also possess an uncertainty relation. We therefore conclude that the measurement of the information range ΔI and the measurement of the causality range Δk are also uncertain. Therefore, we have:

$$\Delta p_x \Delta x \approx h \; (\textbf{\textit{Uncertainty of matter in space}})$$

$$\Delta L_{xy} \Delta \varphi_{xy} \approx h \; (\textbf{\textit{Uncertainty of matter in rotational space}})$$

$$\Delta E \Delta t \approx h \;\; (\textbf{\textit{Uncertainity of energy in time}})$$

$$\Delta I \Delta k \approx h \; (\textbf{\textit{Uncertainty of information in causality}})$$

The uncertainty of information in causality can be formally derived as follows. h is the smallest possible information (see the last paragraph of point 4 above). Hence, ΔI cannot be smaller than h. Hence, Δk has to be the identity element in this case.

The first uncertainty relation is explained by Werner Heisenberg in the following manner. The more accurate the measurement of a spatial position of a particle, the smaller the wavelength of the measuring wave must be. But the smaller the wavelength of the measuring wave is, the bigger the momentum of the measuring wave is, such that the impact on the momentum of the measured particle is big, leading to a big uncertainty of said momentum. The more accurate the measuring of a momentum of a particle, the bigger the wavelength of the measuring wave must be, in order not to influence the particle momentum. But the bigger the wavelength of the measuring wave is, the less precise the measurement of the spatial position of the particle is, leading to a big uncertainty is said position. Hence, momentum and space cannot be determined with big accuracy at the same time.

The second uncertainty relation can be explained in a corresponding manner. The more accurate the measurement of a rotation angle of a particle, the smaller the wavelength parallel to the rotational movement of the measuring wave must be. But the smaller the wavelength parallel to the rotational movement of the measuring wave is, the bigger the angular momentum of the measuring wave is, such that the impact on the angular momentum of the measured particle is big, leading to a big uncertainty of said angular momentum. The more accurate the measuring of an angular momentum of a particle, the bigger the wavelength parallel to the rotational movement of the measuring wave must be, in order not to influence the particle angular momentum. But the bigger the wavelength parallel to the rotational movement of the measuring wave is, the less precise the measurement of the rotation angle of the particle is, leading to a big uncertainty is said rotation angle. Hence, angular momentum and rotation angle cannot be determined with big accuracy at the same time.

The third uncertainty relation is explained by Werner Heisenberg in a similar manner. The more accurate the measurement of a temporal position of a particle, the bigger the frequency of the measuring wave must be. But the bigger the frequency of the measuring wave is, the bigger the energy of the measuring wave is, such that the impact on the energy of the measured particle is big, leading to a big uncertainty of said energy. The more accurate the measuring of an energy of a particle, the smaller the frequency of the measuring wave must be, in order not to influence the particle energy. But the smaller the frequency of the measuring wave is, the less precise the measurement of the temporal position of the particle is, leading to a big uncertainty is said position. Hence, energy and time cannot be determined with big accuracy at the same time.

The fourth uncertainty relation can be explained as follows. The more accurate the measurement of the information of a particle, the more isolated the particle must be, in order to be able to measure its energy, momentum, and angular momentum states. But the more isolated said particle is, the more interactions with other particles are destroyed, leading to a big uncertainty of causality. The more accurate the measurement of causality shall be, the more particle interactions shall be observed. But the more particle interactions shall be observed, the less isolated single particles shall be, leading to a big uncertainty of the information of a single particle. Hence, information and causality cannot be determined with big accuracy at the same time.

Corollary 10: The substances matter (as defined by its momentum and angular momentum), energy, and information lead to an uncertainty relation in their existence forms, space, time, and causality.

Corollary 11: In the physical view of a single local reference frame, there are 8 dimensions: 3 dimensions of space, (x, y, z), 3 complementary dimensions of space given by the overall momenta $((\Delta p_x, \Delta L_{yz}), (\Delta p_y, \Delta L_{zx}), (\Delta p_z, \Delta L_{xy}))$ [1], 1 dimension of time (t), and 1 complementary dimension of time given by the energy (E) [1]. In the physical view of several local reference frames, there are 10 dimensions: the 8 dimensions of the single local reference frame, 1 dimension of causality (k), and 1 complementary dimension of causality given by the information (I).

9. Final remarks on the nature of information

The derivation of information achieved in this paper, and especially the derivation of the law of conservation of information, appear astonishing viewed from the point of view of information technology. Indeed, in information technology we have Shannon's definition of information by using probabilities. Furthermore, information can be always destroyed and created, such that a law of conservation of information makes no sense in information technology.

This apparent discrepancy can be best resolved by analysing the nature of energy in physics and engineering. The law of conservation of energy is a very old principle in physics, and is the basis for many physical derivations. Nevertheless, when performing engineering, like constructing an electric power station, one arrives at the conclusion that not all energy forms are suitable for the set task. Indeed, when converting the mechanical energy of the turbine into electrical energy by the generator, one obtains the equation:

$$E_{mechanical} = E_{electrical} + E_{heat}$$

This equation depicts the thermodynamic principle that the mechanical energy of the turbine cannot be entirely transformed to electrical energy; instead heat is produced during the energy transformation process [1].

Physically spoken, no energy is lost, since the energy before the interaction (the mechanical energy) is exactly the same as the energy after the interaction (the electrical and the heat energy). But from the point of view of an engineer, there is a loss, since the heat energy cannot be used and therefore defines a technological loss. Hence, in engineering, the heat energy is minimized as much as possible. For this reason, an energy efficiency factor η_e is defined in engineering:

$$\eta_e = \frac{E_{electrical}}{E_{mechanical}}$$

This factor is between 0 and 1, and shall be as much as possible near 1, but can never achieve 1 due to thermodynamic principles [1]. This shows that the fundamental physical principle of the conservation of energy is not always of interest in engineering; instead physically equivalent energy types are rated in engineering, defining some energy types as desired and others as not desired, such that energy losses might appear in engineering.

The same reasoning as for energy applies also for information. Indeed, it has been shown in this paper that information is conserved in physics. Hence, from a physical point of view, information is never lost. But from the point of view of information technology, some information pieces $I_1 \dots I_i$ might be desired, since they represent information which can be used for a given purpose, whereas other information pieces $I_{i+1} \dots I_n$ might be undesired, since they cannot be used for said given purpose.

Hence, information pieces are rated in information technology with respect to their usability. Therefore, in information technology an information loss is possible, due to the fact of the presence of desired and undesired information. One could define an information efficiency factor η_i for information technology:

$$\eta_i = \frac{I_{usable}}{I_{overall}}$$

This factor is between 0 and 1, and shall be as much as possible near 1, but can never achieve 1 due to uncertainty principles between information and causality. This shows that the fundamental physical principle of the conservation of information is not always of interest in information technology; instead physically equivalent information pieces are rated in information technology, defining some information pieces as desired and others as not desired, such that information losses might appear in information technology.

Shannon's definition of information by using probabilities is then another form of defining desired and not desired information by weighting some information pieces more than others. This is in line with the point of view of engineering which rates energy, information, and momenta (although momenta have not been addressed in this discussion of point 9). But the ratings performed in engineering do not contradict the physical conservation laws for momentum, angular momentum, energy, and information.

Bibliography

[1]: Carlo Maria Becchi; Massimo D'Elia: Introduction to the Basic Concepts of Modern Physics; Special Relativity, Quantum and Statistical Physics; Third Edition; Springer; 2016.

[2]: https://en.wikipedia.org/wiki/Planck_units#Cosmology

[3]: C. A. Petri: Nets, time and space; Theoretical Computer Science 153; 1996.